Product Patterns with Tens

Example

Find the missing number.

5 × 3 = 15, so 5 × 30 = ■

Think: **5 × 3 = 15**

So, **5 × 30 = 150**.

Find the missing number.

1. 8 × 1 = 8, so 8 × 10 = ■
2. 3 × 3 = 9, so 3 × 30 = ■
3. 2 × 3 = 6, so 2 × 30 = ■
4. 2 × ■ = 18, so 2 × 90 = 180
5. 8 × 3 = ■, so 8 × 30 = 240
6. 6 × 6 = 36, so 6 × 60 = ■
7. 6 × 7 = 42, so ■ × 70 = 420
8. 6 × 4 = 24, so 6 × 40 = ■
9. 5 × 4 = 20, so 5 × 40 = ■
10. 4 × ■ = 32, so 4 × 80 = 320
11. 4 × 9 = ■, so 4 × 90 = 360
12. 7 × 2 = 14, so 7 × ■ = 140

Answer Box

A	B	C	D	E	F
24	200	9	80	360	240
G	**H**	**I**	**J**	**K**	**L**
60	36	6	8	20	90

Objective: Find the product of tens multiplied by a 1-digit number.

Surrounding Patterns

Find the missing number.

1. $3 \times 2 = 6$, so $3 \times 20 =$ ■
2. $2 \times 4 = 8$, so $2 \times 40 =$ ■
3. $3 \times 6 =$ ■, so $3 \times 60 = 180$
4. $9 \times 2 = 18$, so $9 \times 20 =$ ■
5. $9 \times 3 = 27$, so $9 \times$ ■ $= 270$
6. $8 \times 5 = 40$, so $8 \times 50 =$ ■
7. $4 \times$ ■ $= 32$, so $4 \times 80 = 320$
8. $5 \times 6 = 30$, so $5 \times 60 =$ ■
9. $6 \times 2 = 12$, so ■ $\times 20 = 120$
10. $8 \times 7 =$ ■, so $8 \times 70 = 560$
11. $7 \times 4 = 28$, so $7 \times$ ■ $= 280$
12. $7 \times 8 = 56$, so $7 \times 80 =$ ■

Answer Box

A	B	C	D	E	F
30	40	560	300	8	56
G	**H**	**I**	**J**	**K**	**L**
6	18	80	400	60	180

Objective: Find the product of tens multiplied by a 1-digit number.

Product Patterns with Hundreds

Example

Find the missing number.

4 × 3 = 12, so 4 × 300 = ■

Think: **4** × **3** = **12**

So, **4** × **3**00 = **1,2**00.

Find the missing number.

1. 3 × 2 = 6, so 3 × 200 = ■
2. 2 × 2 = 4, so 2 × 200 = ■
3. 2 × 6 = ■, so 2 × 600 = 1,200
4. 6 × 6 = 36, so 6 × 600 = ■
5. 5 × 8 = 40, so 5 × 800 = ■
6. 8 × 2 = 16, so 8 × ■ = 1,600
7. 8 × 4 = 32, so 8 × 400 = ■
8. 9 × 4 = ■, so 9 × 400 = 3,600
9. 6 × 9 = 54, so ■ × 900 = 5,400
10. 3 × 4 = 12, so 3 × 400 = ■
11. 4 × 8 = ■, so 4 × 800 = 3,200
12. 4 × 5 = 20, so 4 × 500 = ■

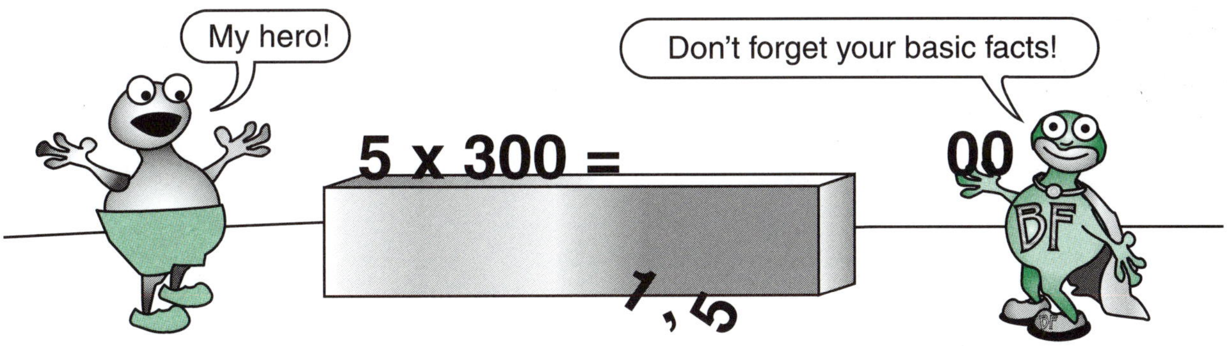

Answer Box

A	B	C	D	E	F
200	36	12	32	3,600	1,200
G	**H**	**I**	**J**	**K**	**L**
2,000	400	4,000	6	3,200	600

Objective: Find the product of hundreds multiplied by a 1-digit number.

Patterns Everywhere

Find the missing number.

1. 4 × 2 = 8, so 4 × 200 = ■
2. 2 × 3 = 6, so 2 × 300 = ■
3. 6 × 4 = 24, so 6 × ■ = 2,400
4. 4 × 4 = ■, so 4 × 400 = 1,600
5. 7 × 1 = 7, so 7 × ■ = 700
6. 6 × ■ = 48, so 6 × 800 = 4,800
7. 5 × 3 = ■, so 5 × 300 = 1,500
8. 5 × 2 = 10, so 5 × 200 = ■
9. 2 × 8 = 16, so 2 × 800 = ■
10. 8 × 5 = 40, so 8 × 500 = ■
11. 6 × 3 = 18, so ■ × 300 = 1,800
12. 3 × 5 = 15, so 3 × 500 = ■

Answer Box

A	B	C	D	E	F
1,500	1,000	15	4,000	100	6
G	**H**	**I**	**J**	**K**	**L**
8	600	16	800	400	1,600

 Objective: Find the product of hundreds multiplied by a 1-digit number.

Multiplication Roundup

Example

Find the best estimate.

7 × 68 Round the 2-digit factor to the nearest ten. 7 × 70 = 490

So, 7 × 68 is about **490**.

Find the best estimate.

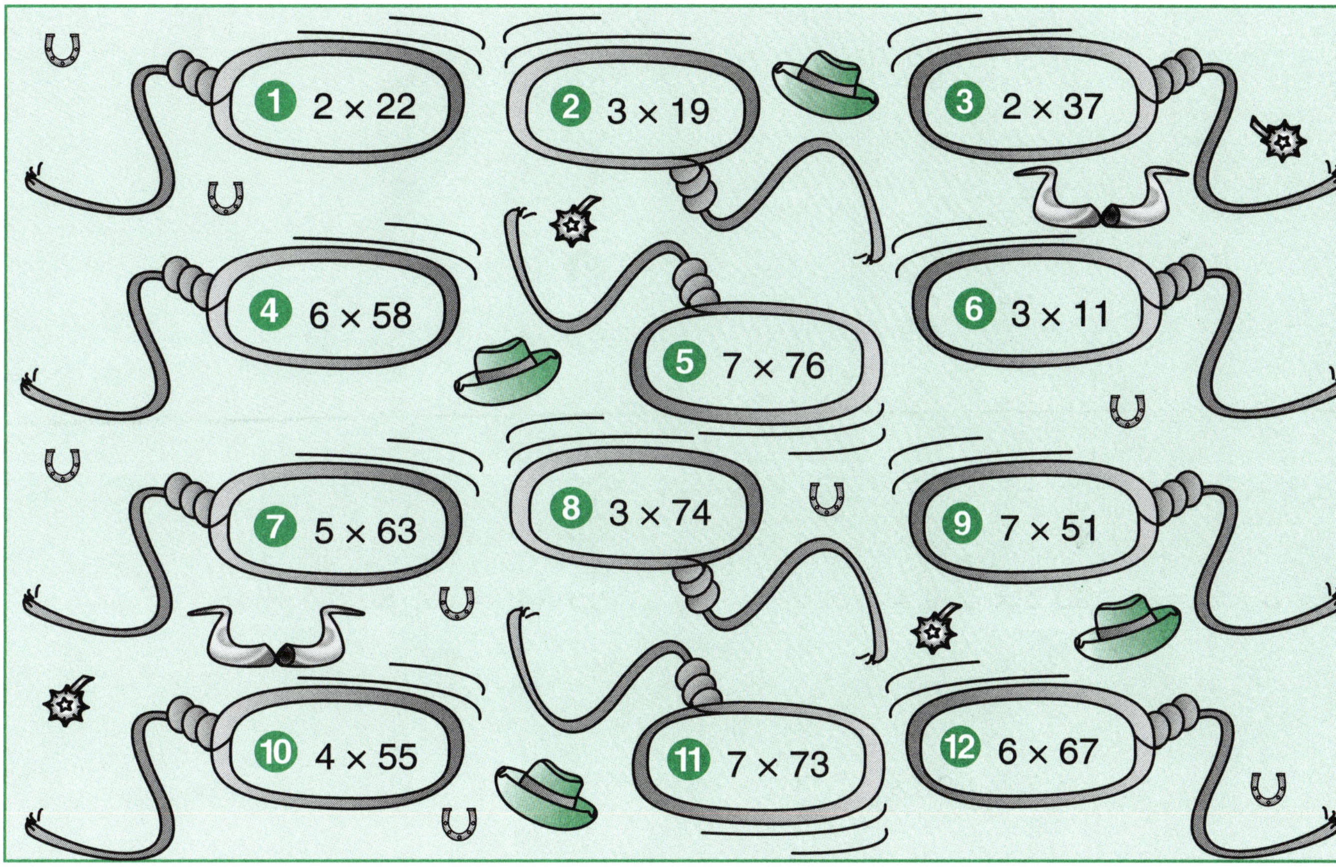

Answer Box

A	B	C	D	E	F
240	360	350	80	60	420
G	**H**	**I**	**J**	**K**	**L**
40	300	560	490	210	30

Objective: Find the estimated product of a 2-digit number multiplied by a 1-digit number.

Getting Around

Example

Find the best estimate.

4 × 316 Round the 3-digit factor to the nearest hundred. 4 × 300 = 1,200
So, 4 × 316 is about **1,200**.

Find the best estimate.

1. 8 × 523
2. 6 × 395
3. 3 × 669
4. 4 × 535
5. 5 × 140
6. 9 × 634
7. 4 × 867
8. 2 × 489
9. 6 × 779
10. 4 × 807
11. 7 × 853
12. 2 × 891

Answer Box

A	B	C	D	E	F
4,000	3,600	1,800	1,000	6,300	2,100
G	**H**	**I**	**J**	**K**	**L**
3,200	2,000	2,400	5,400	4,800	500

Objective: Find the estimated product of a 3-digit number multiplied by a 1-digit number.

Let's Multiply!

Example

Find the product.

$$\begin{array}{rl} 13 & \\ \underline{\times\ 4} & \\ 12 & \text{Multiply the ones.} \quad 4 \times 3 = 12 \\ \underline{+\ 40} & \text{Multiply the tens.} \quad 4 \times 10 = 40 \\ \mathbf{52} & \text{Add.} \end{array}$$

Find the product.

1. 26×2
2. 54×3
3. 26×6
4. 42×4
5. 32×2
6. 21×2
7. 44×8
8. 22×4
9. 67×6
10. 64×8
11. 38×4
12. 48×3

Answer Box

A	B	C	D	E	F
402	156	52	168	162	42
G	**H**	**I**	**J**	**K**	**L**
64	152	352	512	144	88

Objective: Find the product of a 2-digit number multiplied by a 1-digit number.

Ones First, Then Tens

Find the product.

1. 23×4
2. 18×3
3. 57×6
4. 36×9
5. 44×7
6. 27×5
7. 14×4
8. 32×4
9. 29×3
10. Gavin buys 4 packs of pumpkin seeds. There are 37 seeds in a pack. How many seeds does he buy?
11. Morgan plants 3 seeds in each pot. She uses 15 pots. How many seeds does she plant?
12. Patrick plants 24 bulbs in one row. He plants 3 rows. How many bulbs does he plant?

Answer Box

A	B	C	D	E	F
45	308	72	128	56	148
G	**H**	**I**	**J**	**K**	**L**
342	87	135	54	92	324

Objective: Find the product of a 2-digit number multiplied by a 1-digit number.

Another Way to Record

Find the product.

1. 61×3
2. 63×2
3. 22×4
4. 62×3
5. 31×7
6. 81×5
7. 21×9
8. 42×2
9. 51×8
10. 54×2
11. Arlon puts 4 photos on a page. He fills 32 pages. How many photos does he have?
12. Toshi has 72 pages in each photo album. She has 3 albums. How many pages does she have?

Answer Box

A	B	C	D	E	F
405	126	88	408	186	183
G	**H**	**I**	**J**	**K**	**L**
84	216	128	217	108	189

Objective: Find the product of a 2-digit number multiplied by a 1-digit number.

Time to Regroup!

Example

Find the product.

35 × 3 = 15 + 90 = **105**	35 × 3 (regroup 1) = **105**	Multiply the ones. Regroup 15 ones as 1 ten and 5 ones.	3 × 5 = 15
		Multiply the tens.	3 × 30 = 90
		Add 1 more ten.	90 + 10 = 100

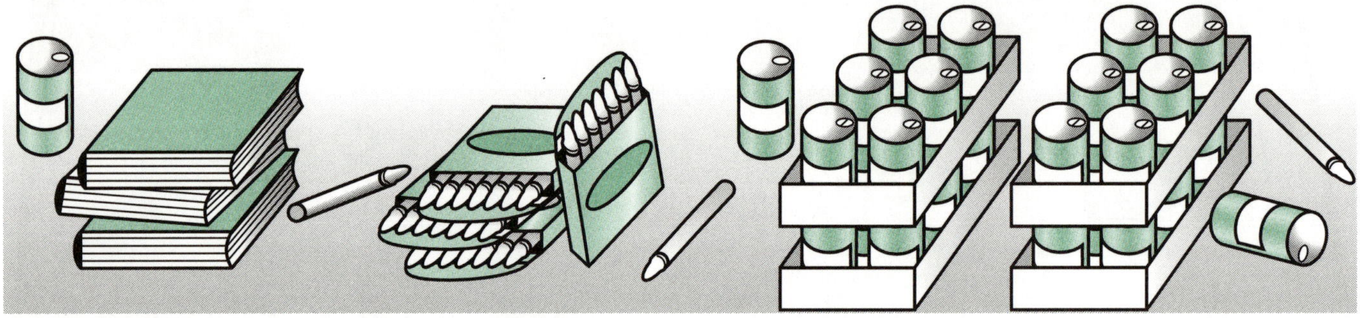

Find the product.

1. 42 × 6
2. 65 × 7
3. 17 × 4
4. 67 × 2
5. 74 × 6
6. 87 × 3
7. 56 × 5
8. 97 × 5
9. 53 × 8
10. 95 × 3
11. Nicole puts 42 shells on each picture frame. She makes 7 frames. How many shells does she use?
12. There are 24 children in Ms. Milton's class. She gives each child 9 crayons for an art project. How many crayons does Ms. Milton need?

Answer Box

A	B	C	D	E	F
216	485	134	252	294	444
G	**H**	**I**	**J**	**K**	**L**
455	261	285	280	68	424

Objective: Find the product of a 2-digit number multiplied by a 1-digit number.

Plenty of Time

Find the product.

1. 73 × 8
2. 59 × 6
3. 38 × 7
4. 89 × 2
5. 42 × 4
6. 89 × 4
7. 39 × 3
8. 72 × 4
9. 86 × 3
10. 94 × 6
11. 38 × 6
12. 14 × 8

Answer Box

A	B	C	D	E	F
178	584	168	228	354	356
G	**H**	**I**	**J**	**K**	**L**
564	117	258	266	112	288

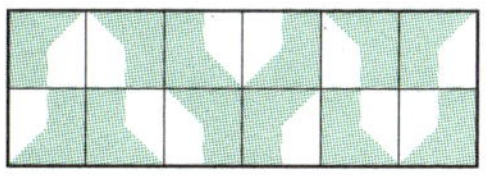

Find the product.

1. 35 × 6
2. 16 × 4
3. 66 × 5
4. 51 × 2
5. 75 × 9
6. 86 × 6
7. 53 × 8
8. 65 × 9
9. 75 × 4
10. 57 × 8
11. 28 × 4
12. 36 × 5

Answer Box

A	B	C	D	E	F
102	210	516	64	330	300
G	**H**	**I**	**J**	**K**	**L**
424	456	180	585	675	112

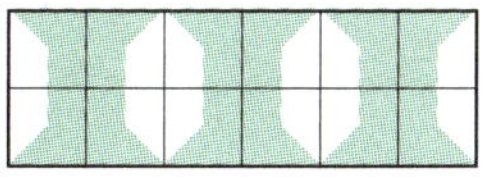

Objective: Find the product of a 2-digit number multiplied by a 1-digit number.

Take a Ride!

Find the product.

1. $\begin{array}{r} 52 \\ \times\ 7 \\ \hline \end{array}$

2. $\begin{array}{r} 36 \\ \times\ 4 \\ \hline \end{array}$

3. $\begin{array}{r} 49 \\ \times\ 8 \\ \hline \end{array}$

4. $\begin{array}{r} 74 \\ \times\ 3 \\ \hline \end{array}$

5. $\begin{array}{r} 69 \\ \times\ 6 \\ \hline \end{array}$

6. $\begin{array}{r} 61 \\ \times\ 6 \\ \hline \end{array}$

7. $\begin{array}{r} 89 \\ \times\ 5 \\ \hline \end{array}$

8. $\begin{array}{r} 53 \\ \times\ 4 \\ \hline \end{array}$

9. $\begin{array}{r} 67 \\ \times\ 2 \\ \hline \end{array}$

10. $\begin{array}{r} 46 \\ \times\ 7 \\ \hline \end{array}$

11. A bus has seats for 78 people. How many people can ride on 5 buses?

12. A total of 64 people can ride in one train car. How many people can ride in 7 cars?

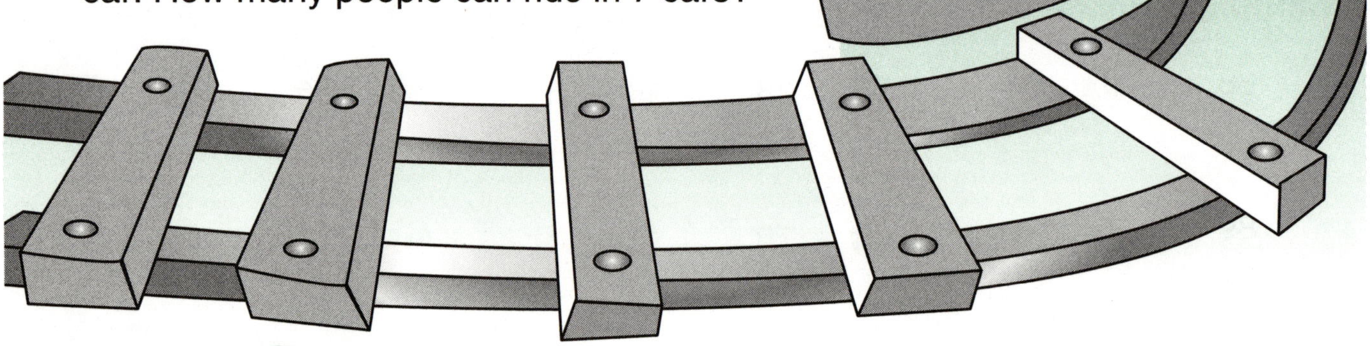

Answer Box

A	B	C	D	E	F
134	392	364	222	366	144
G	**H**	**I**	**J**	**K**	**L**
414	390	322	445	212	448

Objective: Find the product of a 2-digit number multiplied by a 1-digit number.

Give Me Some More Times!

Find the product.

❶ 53×6	❷ 27×5	❸ 48×8	❹ 59×2	❺ 17×8	❻ 62×9
❼ 47×3	❽ 66×8	❾ 35×3	❿ 77×4	⓫ 23×7	⓬ 91×4

Answer Box

A	B	C	D	E	F
136	161	364	135	141	308
G	**H**	**I**	**J**	**K**	**L**
384	105	558	528	118	318

Find the product.

❶ 38×6	❷ 65×8	❸ 41×5	❹ 19×9	❺ 52×4	❻ 45×5
❼ 89×2	❽ 98×4	❾ 84×3	❿ 96×5	⓫ 64×4	⓬ 57×7

Answer Box

A	B	C	D	E	F
392	399	256	205	480	178
G	**H**	**I**	**J**	**K**	**L**
225	520	252	208	171	228

Objective: Find the product of a 2-digit number multiplied by a 1-digit number

Problem Solving: Solving Two-Step Problems

TODAY'S FLIGHTS

Destination	Number of Passengers
New York	148
Miami	423
Cleveland	216
Austin	315
Los Angeles	447
Santa Fe	244
Minneapolis	322

Use the information in the table to solve the problem.

1. The flight to New York stops in Washington. Seventy-nine passengers get off. Forty-six more passengers get on the plane. How many passengers are on the plane now?

2. A flight attendant has 75 chicken dinners and 105 beef dinners. He serves 44 dinners. How many dinners are left?

3. The flight to Santa Fe stops in Chicago. Eighty-six more passengers get on. Then the flight stops in Dallas. Thirty-nine passengers get off. How many people are on the plane now?

4. A flight attendant has 150 cans of soda and 175 cans of juice. She serves 302 cans of drinks. How many cans are left?

5. The plane flying to New York has 37 rows of seats. There are 5 seats in each row. How many seats are empty?

6 The flight to Austin is full. Fifteen adults and nine children still want to get on the plane. How many people want to take this flight?

7 There are 2 pilots and 9 flight attendants on the flight to Minneapolis. How many people will ride on the plane in all?

8 A flight attendant has 117 sports magazines and 220 news magazines for a flight to Los Angeles. How many more magazines are needed for each passenger to have a magazine?

9 Each passenger on the flight to Cleveland gets 1 meal. Twelve meals are packed in a tray. Eight trays have been served. How many passengers still need to be served?

10 A plane has 42 rows of seats. In each row there are 3 seats to the right of the aisle and 2 seats to the left of the aisle. How many seats are there?

11 Eight breakfasts are packed in a tray. A flight attendant serves 15 trays. How many people on the flight to Miami still need to be served breakfast?

12 The flight to Los Angeles stops in Tulsa. Eighty-three passengers get off. Then it stops in Dallas. Fifty-one more passengers get off. How many passengers are still on the plane?

Answer Box

A	B	C	D	E	F
339	136	115	23	291	120
G	**H**	**I**	**J**	**K**	**L**
313	110	210	333	37	303

Objective: Solve a two-step problem that may involve addition, subtraction, or multiplication.

Dollars and More Dollars

Find the product.

1. $36 × 4
2. $75 × 3
3. $27 × 6
4. $83 × 2
5. $28 × 2
6. $83 × 4
7. $43 × 5
8. $67 × 2
9. $52 × 8
10. $62 × 6

11. Each of 3 friends spend $19 for a present. How much do they spend altogether?
12. A train ticket costs $59. Seven friends each buy a ticket. How much do they spend altogether?

Answer Box

A	B	C	D	E	F
$144	$166	$416	$162	$225	$332
G	**H**	**I**	**J**	**K**	**L**
$372	$215	$57	$56	$134	$413

Objective: Find the product of a whole number money amount multiplied by a 1-digit number.

Division Patterns with Tens

Example

Find the missing number.

6 ÷ 3 = 2, so 60 ÷ 3 = ■

Think: **6** ÷ **3** = **2**

So, **6**0 ÷ **3** = **2**0.

Find the missing number.

1. 4 ÷ 2 = 2, so 40 ÷ 2 = ■
2. 6 ÷ 2 = ■, so 60 ÷ 2 = 30
3. 8 ÷ 4 = ■, so 80 ÷ 4 = 20
4. 4 ÷ 4 = 1, so ■ ÷ 4 = 10
5. ■ ÷ 3 = 3, so 90 ÷ 3 = 30
6. 6 ÷ 2 = 3, so 60 ÷ 2 = ■
7. ■ ÷ 3 = 2, so 60 ÷ 3 = 20
8. 8 ÷ 2 = ■, so 80 ÷ 2 = 40
9. 9 ÷ 3 = 3, so ■ ÷ 3 = 30
10. ■ ÷ 4 = 2, so 80 ÷ 4 = 20
11. 6 ÷ 3 = 2, so ■ ÷ 3 = 20
12. 8 ÷ 2 = 4, so ■ ÷ 2 = 40

Answer Box

A	B	C	D	E	F
20	40	80	4	90	2
G	**H**	**I**	**J**	**K**	**L**
8	6	3	30	60	9

Objective: Find the quotient of tens divided by a 1-digit number.

More Patterns with Tens

Find the missing number.

1. 8 ÷ 2 = 4, so 80 ÷ 2 = ■
2. ■ ÷ 2 = 3, so 60 ÷ 2 = 30
3. 4 ÷ 2 = 2, so 40 ÷ 2 = ■
4. ■ ÷ 2 = 4, so 80 ÷ 2 = 40
5. 9 ÷ 3 = 3, so ■ ÷ 3 = 30
6. ■ ÷ 2 = 2, so 40 ÷ 2 = 20
7. 8 ÷ 4 = ■, so 80 ÷ 4 = 20
8. 6 ÷ 2 = 3, so 60 ÷ 2 = ■
9. 6 ÷ ■ = 2, so 60 ÷ 3 = 20
10. 8 ÷ 4 = 2, so ■ ÷ 4 = 20
11. ■ ÷ 3 = 3, so 90 ÷ 3 = 30
12. 6 ÷ 3 = 2, so ■ ÷ 3 = 20

Answer Box

A	B	C	D	E	F
9	90	80	2	30	60
G	**H**	**I**	**J**	**K**	**L**
20	3	8	40	4	6

Objective: Find the quotient of tens divided by a 1-digit number.

Patterns with Greater Numbers

Example

Find the quotient.

14 ÷ 2 = 7, so 140 ÷ 2 = ■

Think: **14** ÷ **2** = **7**

So, **140** ÷ **2** = **70**.

Find the missing number.

1. 18 ÷ 9 = 2, so 180 ÷ 9 = ■
2. 16 ÷ 4 = ■, so 160 ÷ 4 = 40
3. 54 ÷ 9 = 6, so 540 ÷ 9 = ■
4. 27 ÷ 9 = ■, so 270 ÷ 9 = 30
5. 18 ÷ 9 = ■, so 180 ÷ 9 = 20
6. 64 ÷ 8 = ■, so 640 ÷ 8 = 80
7. 27 ÷ 9 = 3, so 270 ÷ 9 = ■
8. 54 ÷ 9 = ■, so 540 ÷ 9 = 60
9. 42 ÷ 6 = 7, so 420 ÷ 6 = ■
10. 16 ÷ 4 = 4, so 160 ÷ 4 = ■
11. 64 ÷ 8 = 8, so 640 ÷ 8 = ■
12. 42 ÷ 6 = ■, so 420 ÷ 6 = 70

Answer Box

A	B	C	D	E	F
60	70	4	8	3	20
G	**H**	**I**	**J**	**K**	**L**
80	2	6	7	40	30

Objective: Find the quotient of hundreds divided by a 1-digit number.

More Patterns with Greater Numbers

Find the quotient.

Math Olympiad

1. 14 ÷ 7
2. 24 ÷ 4
3. 350 ÷ 7
4. 56 ÷ 8
5. 630 ÷ 7
6. 140 ÷ 7
7. 35 ÷ 7
8. 36 ÷ 9
9. 560 ÷ 8
10. 360 ÷ 9
11. 63 ÷ 7
12. 240 ÷ 4

Answer Box

A	B	C	D	E	F
60	4	70	50	5	40
G	**H**	**I**	**J**	**K**	**L**
6	20	90	9	2	7

 Objective: Find the quotient of hundreds divided by a 1-digit number.

D Is for Divide

Example

Find the quotient.

$\begin{array}{r} 16 \\ 3\overline{)48} \\ -3 \\ \hline 18 \\ -18 \\ \hline 0 \end{array}$	Divide the tens. There is one 3 in 4. Multiply. $3 \times 1 = 3$. Subtract. Bring down the 8. Divide the ones. $18 \div 3 = 6$ Multiply. $3 \times 6 = 18$. There is nothing left.

Find the quotient.

1. $2\overline{)26}$
2. $3\overline{)78}$
3. $5\overline{)75}$
4. $4\overline{)88}$
5. $8\overline{)96}$
6. $4\overline{)44}$
7. $6\overline{)96}$
8. $4\overline{)84}$
9. $2\overline{)54}$
10. $7\overline{)98}$
11. $3\overline{)57}$
12. $4\overline{)68}$

Answer Box

A	B	C	D	E	F
21	17	14	16	19	12
G	**H**	**I**	**J**	**K**	**L**
26	11	13	22	27	15

Objective: Find the quotient of a 2-digit number divided by a 1-digit number with no remainder.

Bead Division

Find the quotient.

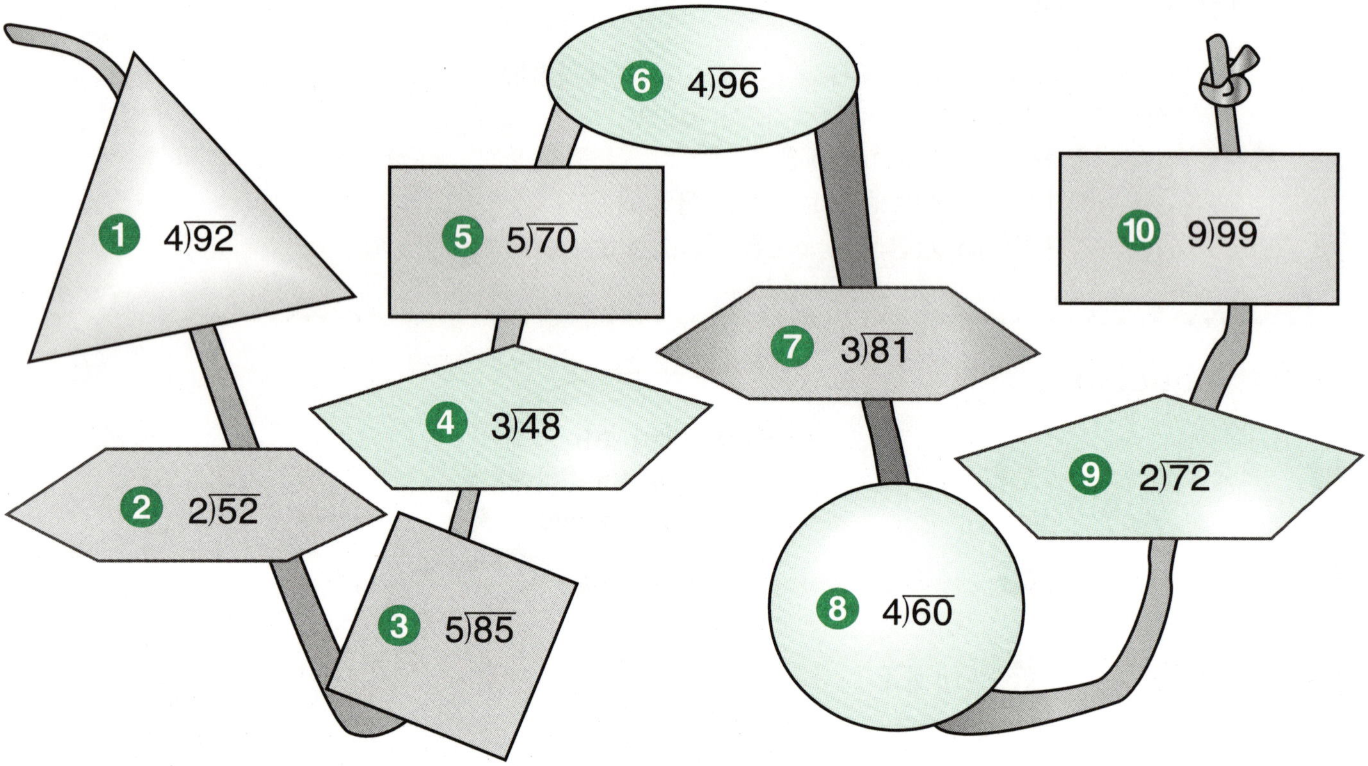

11 Kasa uses 78 beads. She makes 6 necklaces that are the same. How many beads does she put on each necklace?

12 Daniel uses 48 beads. He puts 4 beads on each bracelet. How many bracelets does he make?

Answer Box

A	B	C	D	E	F
23	16	14	36	26	24
G	**H**	**I**	**J**	**K**	**L**
27	11	17	13	12	15

Objective: Find the quotient of a 2-digit number divided by a 1-digit number with no remainder.

Division Cards

Find the quotient.

1. $3\overline{)66}$
2. $8\overline{)96}$
3. $4\overline{)84}$
4. $7\overline{)91}$
5. $3\overline{)96}$
6. $2\overline{)62}$
7. $5\overline{)90}$
8. $4\overline{)68}$
9. $8\overline{)88}$
10. $6\overline{)84}$

11. Lily collected 80 postcards. She put the same number of postcards in 5 albums. How many postcards are in each album?

12. Yori collected 90 postcards. He put 6 postcards on each album page. How many pages did he use?

Answer Box

A	B	C	D	E	F
13	14	12	31	32	11
G	**H**	**I**	**J**	**K**	**L**
18	22	17	15	21	16

Objective: Find the quotient of a 2-digit number divided by a 1-digit number with no remainder.

Divide Them Equally!

Find the quotient.

1. $9\overline{)99}$	2. $2\overline{)58}$	3. $6\overline{)78}$	4. $3\overline{)84}$
5. $4\overline{)64}$	6. $3\overline{)57}$	7. $5\overline{)75}$	8. $2\overline{)64}$
9. $4\overline{)72}$	10. $5\overline{)85}$	11. $3\overline{)93}$	12. $4\overline{)56}$

Answer Box

A	B	C	D	E	F
16	18	17	15	19	32
G	**H**	**I**	**J**	**K**	**L**
13	31	11	28	14	29

Find the quotient.

1. $7\overline{)91}$	2. $2\overline{)52}$	3. $4\overline{)92}$	4. $8\overline{)96}$
5. $3\overline{)75}$	6. $7\overline{)98}$	7. $3\overline{)72}$	8. $2\overline{)34}$
9. $8\overline{)88}$	10. $3\overline{)54}$	11. $6\overline{)90}$	12. $5\overline{)95}$

Answer Box

A	B	C	D	E	F
15	23	26	14	13	24
G	**H**	**I**	**J**	**K**	**L**
11	25	17	19	18	12

Objective: Find the quotient of a 2-digit number divided by a 1-digit number with no remainder.

R Is for Remainder

Example

Find the quotient.

```
  15 R1
 3)46
  -3
   16
  -15
    1
```

Divide the tens. There is one 3 in 4.
Multiply. 3 × 1 = 3. Subtract.
Bring down the 6.
Divide the ones. 16 ÷ 3 = 5 with something left over
Multiply. 3 × 5 = 15. Subtract.
1 is left over. Write it as the remainder.

Find the quotient.

1. 4)75
2. 3)54
3. 7)98
4. 5)74
5. 2)35
6. 5)65
7. 7)95
8. 2)37
9. 4)68
10. 6)80
11. 6)87
12. 3)53

Answer Box

A	B	C	D	E	F
18	17 R2	17 R1	14 R3	13 R2	14 R4
G	**H**	**I**	**J**	**K**	**L**
18 R1	13	14	17	18 R3	13 R4

Objective: Find the quotient of a 2-digit number divided by a 1-digit number with a remainder.

Vegetables for Sale

Find the quotient.

FRESH PRODUCE

1. $7\overline{)91}$

2. $2\overline{)43}$
3. $3\overline{)68}$

4. $5\overline{)69}$
5. $8\overline{)99}$
6. $4\overline{)55}$

7. $6\overline{)77}$
8. $4\overline{)89}$
9. $4\overline{)87}$
10. $2\overline{)44}$

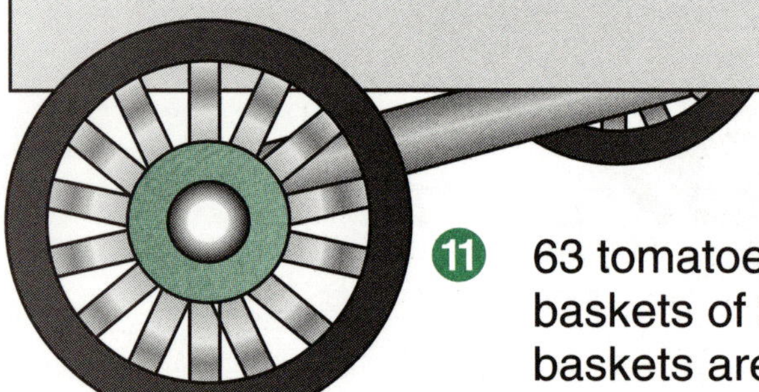

11. 63 tomatoes are packed in baskets of 3. How many baskets are packed?

12. 48 ears of corn are divided equally into 4 bags. How many ears are in each bag?

Answer Box

A	B	C	D	E	F
21 R1	13 R3	22	12 R5	22 R2	21 R3
G	**H**	**I**	**J**	**K**	**L**
12	22 R1	13 R4	13	21	12 R3

Objective: Find the quotient of a 2-digit number divided by a 1-digit number with a remainder.

Leftovers

Find the quotient.

1. $7\overline{)89}$
2. $6\overline{)96}$
3. $5\overline{)57}$
4. $4\overline{)67}$
5. $3\overline{)37}$
6. $5\overline{)84}$
7. $3\overline{)46}$
8. $9\overline{)99}$
9. $5\overline{)77}$
10. $8\overline{)92}$
11. 60 players are divided into teams of 5. How many teams are there?
12. 90 players are divided into 6 equal teams. How many are on each team?

Answer Box

A	B	C	D	E	F
15 R1	11 R4	12	12 R1	15	11
G	**H**	**I**	**J**	**K**	**L**
16 R3	12 R5	15 R2	11 R2	16 R4	16

Objective: Find the quotient of a 2-digit number divided by a 1-digit number with a remainder.

And the Remainder Is . . .

Find the quotient.

1. $5\overline{)94}$
2. $7\overline{)99}$
3. $4\overline{)76}$
4. $4\overline{)59}$
5. $6\overline{)95}$
6. $3\overline{)59}$
7. $4\overline{)74}$
8. $6\overline{)84}$
9. $2\overline{)31}$
10. $5\overline{)99}$
11. A teacher passes out 60 blocks. She gives 4 blocks to each student. How many students get blocks?
12. 36 calculators are divided equally between 2 classrooms. How many calculators does each classroom get?

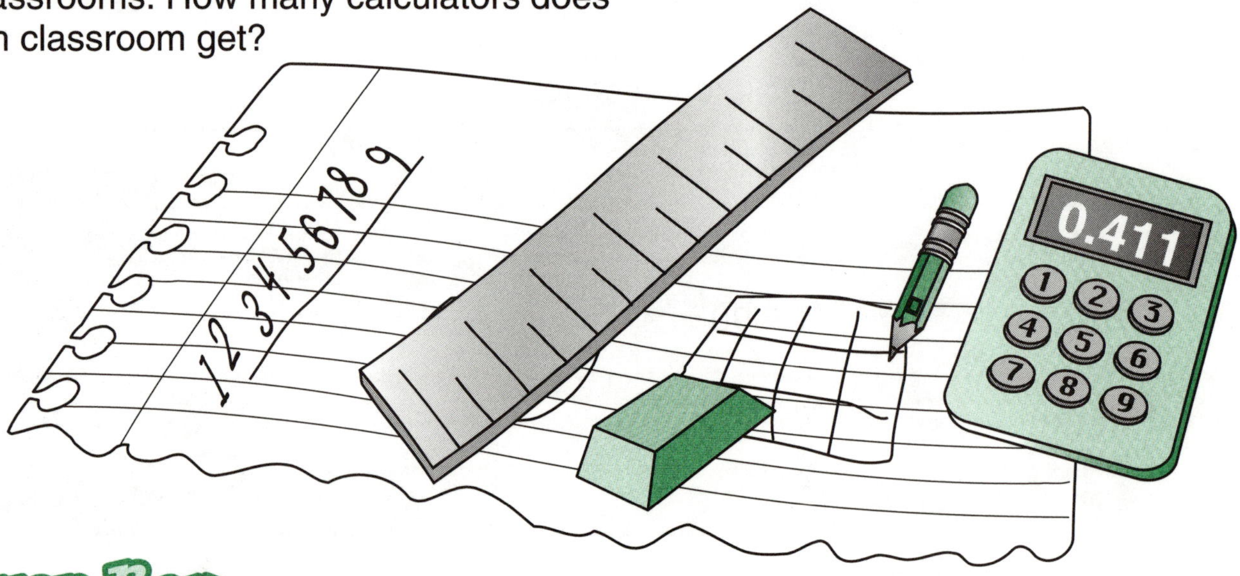

Answer Box

A	B	C	D	E	F
19 R4	18 R2	18	14	15	15 R5
G	**H**	**I**	**J**	**K**	**L**
14 R3	18 R4	19 R2	14 R1	19	15 R1

Objective: Find the quotient of a 2-digit number divided by a 1-digit number with a remainder.

What's Left?

Find the quotient.

1. 4)83
2. 5)82
3. 4)86
4. 5)75
5. 6)97
6. 2)40
7. 6)94
8. 3)62
9. 4)63
10. 2)42
11. 3)48
12. 3)64

Answer Box

A	B	C	D	E	F
15	20 R3	16 R2	21 R1	21 R2	15 R3
G	**H**	**I**	**J**	**K**	**L**
15 R4	21	20 R2	20	16 R1	16

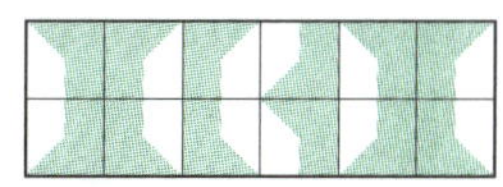

Find the quotient.

1. 7)87
2. 5)90
3. 3)51
4. 6)72
5. 7)96
6. 3)56
7. 5)89
8. 5)68
9. 4)73
10. 6)78
11. 4)70
12. 4)50

Answer Box

A	B	C	D	E	F
13 R5	18 R1	12 R2	18	17 R4	12 R3
G	**H**	**I**	**J**	**K**	**L**
17	17 R2	18 R2	13 R3	12	13

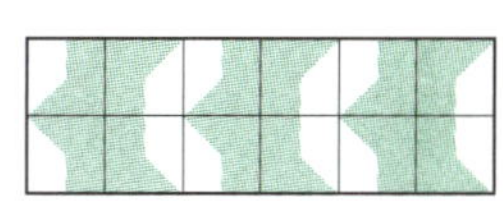

Objective: Find the quotient of a 2-digit number divided by a 1-digit number with a remainder.

Problem Solving: Interpreting the Quotient and Remainder

Item	Beads Needed to Make One Item	Beads Available
Bracelet	6	99
Earring	2	44
Necklace	8	93
Ring	2	75

Use the table to solve the problem.

1. Emily is making earrings. She takes all 44 beads and puts them into groups of two. How many pairs of earrings can Emily make?

2. How many beads does she have leftover?

3. Emily wants to make 20 pairs of earrings. How many beads would be leftover?

4. Debbie has enough beads to make 17 bracelets. True or false?

5. Debbie uses all of the beads available to make bracelets. How many beads does she have left over?

6. She uses a wire for each bracelet. How many wires does she use?

7 Brad uses all of the beads available to make rings. How many rings does he make?

8 After Brad makes his rings, he will have one bead left over. True or false?

9 After Brad makes his rings, he uses any leftover beads to make another ring. How many rings does he make altogether?

10 Rico uses all of the beads available to make necklaces. How many beads will he have leftover?

11 After Rico made his necklaces, he put each necklace in a box. He put any leftover beads in another box. How many boxes did he need?

12 How many necklaces did he make?

Answer Box

A	B	C	D	E	F
16	0	12	3	False	22
G	**H**	**I**	**J**	**K**	**L**
True	11	38	4	5	37

Objective: Decide if the correct answer to a problem is the resulting quotient, the remainder, or the next greater quotient.

Dollar Division

Find the quotient.

1. 4)$28
2. 7)$91
3. 4)$40
4. 5)$70
5. 9)$72
6. 3)$57
7. 2)$30
8. 6)$54
9. 7)$77
10. 4)$68
11. The total cost of 4 video tapes is $64. Each tape costs the same amount. What is the cost of one tape?
12. A tape recorder costs $48. 4 friends decide to split the cost of it equally. How much does each person pay?

Answer Box

A	B	C	D	E	F
$19	$13	$15	$17	$11	$10
G	**H**	**I**	**J**	**K**	**L**
$9	$12	$14	$7	$8	$16

Objective: Find the quotient of a whole-number money amount divided by a 1-digit number with no remainder.